Beginner's Guide to the Colonization of Mars

I0461781

Beginner's Guide to the Colonizations of Mars

Beginner's Guide to the Colonization of Mars

By Jiho Min
Edited by Matt Leuchtmann

Table of Contents

This book is dedicated to my parents, tutors teachers, friends and everybody who always supported me. Special thanks goes to Matt Leuchtmann, who edited my manuscript.

Beginner's Guide to the Colonizations of Mars

Introduction

Mars is considered as one of our most important future destinations . Why? Because we need to colonize Mars in order to extend the human civilization into the outer solar systems and into deep space. Currently, we are considering colonizing the red planet. Many space agencies and private companies are working to send humans to Mars in following decades with the dream of colonizing the planet.

So why should we colonize Mars which could potentially cost billions of dollars? There is no one exact answers but there is surely lots of variety of answers. Some of them are: outpost for further exploration into outer solar system, commercial use such as used in space tourism, being a multiplanetary species, industrial use, scientific research at Mars, and Mars could be a back up plan for us when something bad happens to the Earth. These reasons are all important and surely and greatly will contribute to us.

Many private companies are competing to send astronauts to Mars and colonize the planet. For example, Mars One project is working toward sending astronauts to Mars. Elon Musk of SpaceX said he established his company SpaceX to colonize Mars. Blue Origin, a company supported by billionaire Jeff Bezos, also announced they will colonize Mars in following years. All of the competitors know the real value of Mars. The planet is full of industrial opportunities like mining minerals and being a popular tourist attraction. In this book, I will talk about how can we colonize Mars to maximise benefits and minimize costs.

It is good to know what transportation systems national agencies and private companies that are racing toward Mars's colonization are developing. NASA is developing the Space Launch System (SLS) and Orion capsule and a ship called Deep Space Transport for colonization. SpaceX is developing an entirely new rocket and module called Interplanetary Transport System and Big Falcon Rocket which will be powered by 31 Raptor Rocket Engines. However, this may not be a good idea because Russian engineers had tried to use a rocket with tens of engines and they failed. The Blue Origin is saying that their company will be sending humans to Mars in a few decades. Created in 2011, Mars One project is widely known and unlike any other agencies and companies, they are planning to buy everything that they need in order to send astronauts and colonize Mars.

Colonizing Mars might be really hard and there are lots of dangers that faces us. But, we need to explore Mars because we are the species that explore and figure out anything, whatever it takes.

Ways to Colonize Mars

There are currently two options to build a permanent settlement on the red planet. One method is called terraforming, and the other method is called paraterraforming. Despite their similarities in their name, they are very different.

We will start with more widely known terraforming. Terraforming requires to heat up the entire planet and make Mars, more Earth like.

On the other hand, paraterraforming is to build a covered habitable places to live on Mars.

There are many ways that we can terraform Mars. First way is to use an enormous space mirror. It would reflect sunlight to heat up Mars. Because Mars's atmosphere is mainly composed of carbon dioxide, it will create a greenhouse effect. So, the greenhouse effect would melt the polar cap which will made Mars more Earth Like. This is the whole point of terraforming. Engineers and scientists predict that the largest mirror would have a radius of 125 km and will operate at the 214,000 km above the Martian surface. They predict it will create up to 4×10^{15} Joules of heat.

Another way is to use small Delta V asteroids (up to 300 meters in diameter) suggested by Robert Zubrin. Asteroids consume lots of ammonia inside them. When an asteroid make an impact on Mars, ammonia will be released. Ammonia will spread into the Martian atmosphere and create stronger greenhouse effect which took pretty long time which was about a century. Also, it would increase the pressure on Mars which is currently 1/100th of the Earth's. Robert Zubrin, president of the Mars Society, says optimistically said we will need only 40 asteroids. He said each asteroid will generate up to $3×10^{20}$ Joules and each asteroid would raise up to 3 degrees Celsius.

Although, I disagree with his plan. One reason is because it will destroy the ancient landscape of Mars We could study Mars closely. When we destroy the landscape, which I don't agree, because we are losing all the important scientific resources and it might accidentally hit the existing settlement. Also, humans couldn't breath ammonia. But, mostly, the problem is that currently we don't have ways to move such a big asteroids! This idea might be possible in near future, but not now.

If we put this all together it would be like: First, we heat up the planet. Then, people would move in to create the self sustainable colonies that would sustain itself. Plants breathe in carbon dioxide, so we can use carbon dioxide to fuel the plants, which will provide us with oxygen and food for the colony.

Next, we will going to learn more "efficient" way to colonize Mars which is paraterraforming.

About few years ago almost all scientists had not known what paraterraforming is. But now we now know that some theorists suggest to colonize Mars, we could possible use Lattice Assisted Nuclear Reaction which is abbreviated "LANR". In the issue 131 of Infinite Energy, Jet Energy explained they will use pair of deteriums (heavy hydrogen) to fuse them together to create Helium nucleus. During this process, they will create a large amount of energy with a lot of unexpected heat (known as excess heat) and oxygen which will support the martian colonies. But in very hard-to-achieve conditions.

Now, I am going to explain about our machine (We haven't named it yet). The machine would transform carbon dioxide, which is in the Martian atmosphere, to the oxygen. First, using a giant fan, machine will collect carbon dioxide. Then, we will freeze each molecule to solid state to avoid thermodynamic explosion while we split the molecules. Then, we will split the molecules into 2 oxygen molecules and 1 carbon molecule. Finally, the atom get final check. Carbon will go to the carbon storage and the oxygen will be send to the martian colonies, supplying oxygens.

Although our machine need some brand new technology. First, because we need to split atoms, we will need a nanotechnology. Also, 3D printing is needed to build a complex and big structure on the surface of Mars. Technically in my opinion, all the structures of colonies on Mars will be built by the 3D printing.

However, there is another way to separate carbon from the oxygen. We can use chemicals to separate the molecules. The formula that I had created is: $CO_2 + H_4 \rightarrow CH_4 + O_2$ This meant by using 4 hydrogen compound and carbon dioxide, we can produce methane and oxygen. Methane can be used in different fuel cells and batteries which makes it more efficient.

To Mars!

So we figured out how we could colonize Mars into habitable places on Mars. But how are we transfer that much supplies that we need? Currently, there is a lot of research going on to develop more find the perfect course to Mars.

The opportunity to launch a spaceship to Mars came about every 2 years. So we must send the supplies first and then for the interval about 1 or 2 months. It is because when we send it 2 years before it would be covered in dust and may not be very well functioning.

But, first we need to send supply modules first. The reason behind this is that we need to create a base that would support the astronauts who arrived their. Mars One intends to send six supply modules first and then will send astronauts. The base would contain life supporting unit which include oxygen and water. We also need a solar panels for energy source and a transportation which might be a big rover. In addition we might need a 3D printer because then we could fix things and make stuff.

There is a lot of options to get to Mars. For the very first "era" when we explore Mars, the rocket would first launch from Earth. Then for the fast velocity, we might use the gravitational pull of the Moon. Then, we would go to Mars.

We might need a station with a big firepower. I prefer using a big "ship". The reason is that way to Mars (one-way) took about 6 months. We can't live 6 months in the tiny cramped space. It would be mentally and physically hard for the astronauts. So the station must contain basically almost same things as base on Mars. The problem is that, how we are going to get the strong firepower that could transfer station size of ISS to Mars? We might need an engine stronger engine than Saturn V and SLS.

So, NASA and Boeing developed a plan and announced it in 2017. First, Orion capsule would launch from Earth in SLS. The orion capsule would transport up to 7 astronauts and dock to the capsule called Deep Space Gateway. It looks like the International Space Station but smaller. It would provide plenty of space for astronauts. Gateway will be assembled in space and orbit the Moon. Then the Deep Space Gateway would reach Mars and make the historical landing.

I will explain more about another way to get to and land on Mars: Using moons. It is crucial to use moons because they can help us in many ways!

Stops Before Landing on Mars

It might be good for us to have some interplanetary "stops" along the way before Mars. First, here are some candidates: Moon, Phobos, and Deimos. Actually if we launch rockets on Moon, it would be great because Moon has less gravity than Earth. But, Moon isn't colonized yet. So, in the following decades we must launch rockets from Earth.

If we consider that Moon is our stop, it would have advantages and disadvantages. The advantage is that we could transfer supplies that ship very easily to the Moon. The disadvantage is that it could be hard to ship to land and then take off again which would take lots of fuels.

Phobos and Deimos is in different situation unlike Moon it is the moon that orbits Mars. the plan is that this small moons of Mars could be very useful. It is because they are the "final" stop before touchdown.

There are two types of base that we can make. The first one is built on the moons and another option is to build a base *inside* the asteroid. Yeah, it's a pretty strange idea.

To build a base on the surface, we must deliver the supplies that we might need and either humans or robots that can build bases. Or we might send the "fleet" of ships first and then let ships assemble the base. One useful tool would be 3D printers. We can use 3D printers to build parts that we need and connect them.

If we wished to build the underground base, it would be hard because we have to dig into the moon's surface. However it has some great advantages. Some of the examples are: block the harmful rays from deep space and sun.

Although it has lots of advantages, building an underground base have some serious problems. First, it is really hard to build. It is because we have to dig up the inside of the asteroid. During the process, we might need to use explosives and we might accidentally damage the moon. Also, at least we must build a temporary "building". There is problems that spacecraft that are arriving at the moon. It might be very hard for "space" pilots to enter and exit the underground base.

In my opinion, it might be better if we had a landing craft *at* the Phobos and Deimos. The actual ship would only need to carry people and supplies. The landing craft will be at the moons of Mars. So, this would reduce the amount of the supplies that we need. However we need to carry landing capsules first. It is the thing to consider about.

The Martian Habitat or Base

So, we had our ships arriving at Mars with either LANR or our machine, ready to supply oxygens and build a colony. Wait! How will the settlements (Martian habitat), will look like? That is a very important question to answer. It is because it needs to include all the things that we will need to survive.

First, we need to list things that keep us alive. It is water, food, and air.

If we can prove that there is an underground aquifer on Mars, we might not need to bring much water. But if we don't, we can land closer to poles and get water using pipes. For food, we can use a more advanced greenhouse called MarsFarm. It makes a condition that plant can live in. For right atmosphere for colonists, you can see a section in this book called Ways to Colonize Mars and in paraterraforming part.

The next thing problem to think about the things we need when we need to do something. Such as medical equipment, rover, solar panels for energy…

So the main buildings might be surrounded by domes or partially underground. It is important to have your base partially underground because they block harmful UV rays from passing through and generates heat which can keep the astronauts warm.

Agencies and private companies all considered using solar panels. So, they will be arranged in rows, collecting energies. We also need 3D printers in case colonists needs to build or fix some parts. Rover is needed to travel some distances on Mars.

For the main parts of the building, we might reuse the landing vehicle. It is useful because if we don't reuse the landing vehicle, we would need more launches to deliver structure of the building which is more costly.

Lastly, we might need LENR and maybe our machine. But, we actually didn't need to bring them all the way to Mars because we can produce them using the 3D printers.

Thinking and actually setting up base is very difficult and hard work. But, I am confident that Martian habitats will do their job.

Water on Mars?...

By now, the only known water source on Mars is small caps of frozen water on the polar caps. We also know that the polar cap expand and recede due to the change in the season. Only 85 percent as much ice found in greenland can be found on Mars which is about 800,000 cubic miles. Although, the ice is not perfectly "pure". It is because it is mixed with martian dirt. We want water because it is important to the human body and it is impossible for us to send tons of water for us to use for years on the colonies. So, we need to find water source on Mars.

Over thousands of millennia, moon had stabilized Earth's axis. But, Mars doesn't have a such a huge satellite (Moon is the fifth largest in the Solar System). Its tilt swings 30 degrees in every 100,000 years. The swing of tilt has caused Mars's climate to repeat from milder to active climate. Because its tilt swings, Mars's polar cap expands and recedes over martian "seasons".

At night, the atmosphere of Mars is filled with water vapor. When it is all combined, it would create 10-50 micron deep cover on the surface of Mars if they are all combined. Viking 2 had first observed appearing and disappearing of the frost. In addition fog also had observed.

Mars might contain vast ocean of water for the first half billion years after it had formed. Oceans have formed and dried out. It repeated the process a few times.

I also figured out that aquifer could be formed under the Martian soil. It might as well have underground rivers and oceans which might contain a living organism. They could be lots of water because, mars was once covered by vast oceans. Some think over 20 percent of the surface covered by ocean. We will need to have mining operation to pull water out from the aquifer

Water on Mars is very important for two reasons: 1. People in the martian colonies need to drink water to survive. 2. We need water to water the plants in the future martian vegetation. So this is why water on Mars is especially important.

To provide water for our astronauts, we needed to "mine" the frozen water in the polar caps. Or as I mentioned before we might land at the exact spot near the aquifer and mine water from there. Also we might deploy a water capturer to catch water vapors and get water from there.

Dangers Along the Way

During our way to Mars and build a permanent settlement has lots and lots of problems. First danger that we are going to talk about is astronauts health during their six month journey to the fourth rock from the Sun.

There are two types of health. Mental and physical. Let's talk about physical health first. When we are in microgravity for a long time, our bones will lose calciums precisely 1 percent per month and blood will flow in a different way which is confusing for our body that is suitable in the Earth's 1g. So, this is why the astronauts at ISS always do lots of exercise.

Another problem to consider is the cosmic rays (or radiations) and Ultraviolet light . They are the radiation that was created after the Big Bang and produced from Sun. They could harm our cells. This is what NASA's engineers and scientists who are now building new Orion capsule are considering that problem now. They are developing a good cover for capsule so that radiation couldn't penetrate through.

Mental health also could affect astronauts. Because we are used to live in such a big world and compare to that tiny capsule of Orion and Deep Space Gateway. So brilliant engineers always come up with solutions. In Orion capsule, they attached an inflatable capsule that will inflate after capsule reached space. It will give more space for supplies and astronauts to move around

After arrive on Mars, another problem arose: sandstorm. On Mars, sandstorm is quite common. Also there are planet-sized sandstorm sweeps the entire planet. Sandstorm is very dangerous, because they will knock down the communications and block out all the sunlight so that solar panels didn't work. Quick! Take cover!

During the six-month journey to Mars, there is a very high chance that something might happen to spacecraft. Whether it's malfunction or it is an explosion. I am considering this because the system's are very complicated. Astronauts may fix parts but it is still very dangerous. So, I suggest having a deployment part for emergency.

Energy Source on Mars

All the things in the world are powered by energy. Beyond our planet, on Mars, will be powered by energies. There are many options on Mars to fuel our colonies. We will start with the topic of using a thermal energy.

The thermal energy is from the heat that is escaping Mars's interior. These heat are likely to trapped inside Mars when it was formed 4.6 billion years ago. With Insight's probe called PH3, we can expect how much heat is escaping mars's interior and if is it enough to run the entire colony or not.

Wind energy would be very useful on Mars. Mars is windy place. The strong wind are common and occasionally big sandstorm sweeps the entire planet. With a lot of wind, we might generate thousands of kilowatts that would power the colonies on Mars. But, if the wind is really strong, they might be destroyed. It is a priority to consider.

Like in most spacecraft we could use solar energy provided by solar panels. It would be one of the best choices because we are familiar with them and Mars has thinner atmosphere than Earth so that more sunlight could penetrate through. But, there is a problem. Sandstorm, which is common on Mars, could block out all the sunlight so that solar panels wouldn't work. However most companies and agencies are all considering using the solar panels because we are very familiar with using them. SpaceX and NASA are both announced that they will use solar panel on Mars.

The last energy source that we will talk about is the energy that almost all of our house's electricity is from. It is the nuclear power. The good thing about nuclear power is that they create a lot of energy and also create heat in the process of atom's fission and fusion. But, it might be dangerous because it can damage the martian atmosphere and land.

There are many options that we can make. It is left to us to choose the best option.

Costs and Benefits of Colonization

There are many benefits when we colonize Mars. I refer Mars as Industrial planet. Mainly because Mars is great for industry.

Red planet contain numerous types of minerals including iron which is all over the entire planet. It could has mining operation to drill minerals.

Another benefit that Mars have is it's a great place to launch a spaceship. There are two reasons. First, Mars have weaker gravity than Earth. So, spaceships will have less escape velocity. Also, Mars is closer to the outer Solar system. So Mars is superb place for launching a spaceship. Like in example, we could send a scientific probe to the outer solar system. In addition, we could send asteroid miners.

In addition, Mars could be a tourist attraction. Now people are starting to think space travel as like their travel, like going to space center. So, I'm sure lots of people wanted to visit Mars. And it is true! By the report currently more than 200,000 people had signed up to the one- way ticket to the red planet.

Happy news for scientist! Mars contained a lot of things that scientists have to study. In example geologist can study compositions of rocks on Mars. Biologist can study the evidence of the life that once existed on Mars.

Lastly, we could be multiplanetary species. Being interplanetary species is important because that raise our chance of survival. Think, what should we do when an asteroid the size of the one that killed all the dinosaurs hit the Earth? We must leave the Earth, right? So, we need a back up plan in case of emergency and Mars is our back up plan.

But, all of these things can't come out with 0 dollar. We need money!!! It would even take billions of dollars human and supplies that need to survive during their journey and on the surface of Mars. So, I think the whole entire world should work together. It would be like funding of ISS. This is because nobody has millions and billions of dollars. In the series of film of Mars, the nations came together to form International Mars Science Foundation. It would contain a lot of countries.

Let's hope that we can earn all those benefits when we do colonize Mars..

Conclusion

We talked about various things about Mars. From space mirror to nuclear power source on Mars, we have come to now where we are researching more deeply into the topic of colonization of Mars and actually prepare for reaching our future destination.

We still have lots of problems to solve and try to find a better solution. Such problems include the spaceship big enough to support the crew members which is unbuildable on Earth using the current technology and needs to be built in space and the astronomical cost and also include finding ways to safely deliver astronauts to Mars and lots of other problems.

However, I certainly predict that we will colonize Mars soon. I know that it will be hard. But, the most important thing is to never give up. That's how we will colonize Mars.

Appendix: Colonization of Venus

Colonization of Venus seemed impossible to you who had learn how hard it is to build colonies on barren land because of its unbelievable temperature and extreme pressure. But, we can certainly colonize Venus. However, we need to build the colonies not on its surface, but on top of its carbon dioxide-rich clouds specifically about 31 miles above the surface.

On top of the clouds, the atmospheric pressure would be similar to those in the Earth's sea level and its temperature would be lower at the top of the clouds (0-50 degrees Celsius). So, future inhabitants of Venus might be able to wander around without their space suit if they had the right atmosphere and protection which can be solved by creating the geodesic domes.

Each colony would be the size of the entire city and be floating above the clouds using its buoyancy. We would need lifting gases to increase the buoyancy. These lifting gases would be hydrogen and helium and other cheap and light gases.

For problems with energies, solar panels would work because Venus gets way more sunlight than Earth.

But, there are some huge problems. One problem is that the colonies will be unstable due to Venus's very strong winds that we call the "Super Rotation", which can possibly damage the floating colonies.

Venus might be more unpleasant than Mars, but why not colonize Venus when we colonize Mars?

a. Spaceanswers.com - Future Tech: Floating
colonies on Venus

Bibliography

David A. Weintraub, 2018, Life on Mars: What to
Know Before We Go (Princeton Press)

Craig Devine, 2015, The Colonization of Mars:
From Earth to New Worlds (Westland House
Publishers)

Mitchell Swartz, Theodore Schuster, Gayle Verner,
Joshua Gyllinsky, Jeffery Tolleson, "Paraterraforming
Mars I. Heat, Electricity and Oxygen Are Available
from Lattice-Assisted Nuclear Reaction", 2017, *Infinite
Energy*

Michio Kaku, 2018, The Future of Humanity:
Terraforming Mars, Interstellar Travel, Immortality,
and Our Destination Beyond Earth (Doubleday)

www.ingramcontent.com/pod-product-compliance
Lightning Source LLC
Chambersburg PA
CBHW061237180526
45170CB00003B/1331